Augustin Niyomukiza

MUSURURU beverage; manufacturing analysis and quality improvement

Augustin Niyomukiza

MUSURURU beverage; manufacturing analysis and quality improvement

ScienciaScripts

Imprint

Any brand names and product names mentioned in this book are subject to trademark, brand or patent protection and are trademarks or registered trademarks of their respective holders. The use of brand names, product names, common names, trade names, product descriptions etc. even without a particular marking in this work is in no way to be construed to mean that such names may be regarded as unrestricted in respect of trademark and brand protection legislation and could thus be used by anyone.

Cover image: www.ingimage.com

This book is a translation from the original published under ISBN 978-620-6-71911-3.

Publisher:
Sciencia Scripts
is a trademark of
Dodo Books Indian Ocean Ltd. and OmniScriptum S.R.L publishing group

120 High Road, East Finchley, London, N2 9ED, United Kingdom
Str. Armeneasca 28/1, office 1, Chisinau MD-2012, Republic of Moldova, Europe
Printed at: see last page
ISBN: 978-620-7-93955-8

Table of contents

SUMMARY

Our research is part of the development of the alcoholic beverage "*musururu*" produced locally in the town of Goma. This is why physico-chemical and microbiological analyses of this drink were carried out in order to improve it from a technological, microbiological and health point of view. The semi-structured survey enabled us to identify three different "*musururu*" production technologies in the city of Goma. The critical points in each unit operation in the production of this drink were identified. An improved technology using selected *Saccharomyce cerevisiae* yeast followed by pasteurisation was then proposed. The production of '*musururu*' differs from modern processes in that the brewing operation is more complex, the alcoholic fermentation by the traditional ferment is quicker and shorter, and a natural lactic fermentation is superimposed. A better stability of physico-chemical parameters was observed for samples obtained with the addition of sucrose and *S. cerevisiae* yeast, then stabilised by pasteurisation. For some microbiological parameters, these samples showed variation. A very high variability of certain physico-chemical and microbiological parameters was observed for the sample obtained from local manufacturers.

Key words: Traditional "*musururu*" beer, wild yeast, saccharomyces cerevisiae, alcoholic fermentation, pasteurisation, improvement, preservation.

ABSTRACT

Our research is part of the valorization of the alcoholic drink *"musururu"* produced locally in the city of Goma. This is why the physicochemical and microbiological analyzes of this drink have been carried out to improve it from the technological, microbiological and sanitary point of view. The semi-structured survey enabled us to identify three different technologies for the production of *"musururu"* in the city of Goma. The critical points in each unit operation of the production of this drink have been identified. Then improved technology by using the selected yeast *Saccharomyces cerevisiae* followed by pasteurization was proposed. The production of *"musururu"* differed from modern processes by the greater complexity of the brewing operation, the rapidity and shortness of the alcoholic fermentation by the traditional ferment, superimposed on a natural lactic fermentation. A better physico-chemical parameters stability was observed for the samples obtained with the addition of sucrose, yeast, *S. cerevisiae* and then stabilized by pasteurization. For some microbiological parameters, these samples showed a variation. A great variability of certain physicochemical and microbiological parameters was observed for the samples obtained from the local manufacturers.

Keywords: Traditional *"musururu"* beer, indigenous yeast, saccharomyces cerevisiae, alcoholic fermentation, pasteurisation, improvement, preservation.

INTRODUCTION

Common sorghum (*Sorghum bicolor*) is an annual cereal of tropical origin, belonging to the grass family. It is grown in semi-arid areas of Africa and Asia (Louise et al, 2007). It can be grown in all types of soil (dry, waterlogged or highly saline) and is highly resistant to hot and dry agro-ecologies. These qualities give it a considerable advantage over other cereals (François, 2013). It is the world's fifth most important cereal in terms of production, after wheat, rice, maize and barley, with 55 to 60 million tonnes produced annually in 2010, including 25.6 million tonnes in Africa (Bwanganga et al, 2013).

Sorghum grains can be eaten whole or shelled to make porridge, couscous or pancakes. Sorghum is a major source of energy and protein for millions of people in Asia and Africa. It is one of the main foodstuffs in the poorest regions of the world, where food security is threatened (FAO, 1990). It is used in brewing as a complement or substitute for barley malt. Although its use is recent, it is increasingly used in the production of gluten-free beer (Dicko et al 2005).

In Africa, sorghum is often malted and transformed into a traditional drink, the production of which includes an essential alcoholic fermentation stage (Aka et al, 2008). The production process and the name of the drink vary from country to country and region to region: *TCHAPOLO* in Côte d'Ivoire, *IKIGAGE* in Rwanda, *IMPEKE* in Burundi, *DOLO* in Burkina Faso and *UTSHWALA* in South Africa (Coulibaly, 2013). In the east of the Democratic Republic of Congo, specifically in the province of North Kivu, this drink is known as "*MUSURURU*".

Musururu" has a remarkable socio-economic character, and is served at various ceremonies and celebrations (weddings, funerals, dowries, ...). The drink is a significant source of income for small-scale producers (especially women), who produce it locally using traditional technology. The relative success of this drink with consumers is justified by the fact that its price is relatively low compared with western lager-type beers made by two firms (Bralima and Brasimba) based in DR. Congo (François, 2013). Its high calorie content, B vitamins including thiamine, folic acid, riboflavin and nicotinic acid, and essential amino acids such as lysine make it a nutritional supplement (Maoura et al, 2007).

The empirical technological process involves a double fermentation: alcoholic fermentation combined with natural lactic fermentation (Aka et al, 1998). Production conditions have not changed over time. The sprouted sorghum grains are still dried in the open air, along paths frequented by flies and insects. What's more, the production process suffers from a crucial lack of precision measuring instruments and good hygiene practices, and the successive sweet musts are inoculated with the ferment from the previous fermentation, with little knowledge of the actual nature of this ferment (Aka et al 1998). The resulting beverages are unstable, of inconsistent quality and deteriorate during storage, resulting in a loss of earnings for producers. This is why it is necessary to assess the microbiological and physico-chemical quality of this beverage and think about improving it from a technological, microbiological and health point of view.

Our hypothesis was that the microbiological quality of this drink is poor and exposes consumers to health risks. It also leads to huge losses during storage. Controlling the critical points in the manufacturing process and developing a new product obtained using a selected yeast (*S. cerevisiae*), then stabilised by pasteurisation, would improve the microbiological and physico-chemical quality of this drink.

The aim of this research is to improve the quality of home-made "*musururu*". Specifically, the research aims to:

- ✓ Describe the production diagram for the "*musururu*" drink;
- ✓ Identify critical points on the production diagram and improve them,
- ✓ Analysing the microbiological and physico-chemical parameters of traditional and improved drinks.

The research aims to gather together all the information relating to the manufacturing technology and hygienic quality of the traditional "*musururu*" drink. It then proposes a new technology to improve the microbiological and physico-chemical quality of the drink. This information can be used for the benefit of manufacturers and the health of consumers of such drinks.

CHAPTER 1. GENERAL INFORMATION ON SORGHUM, TRADITIONAL SORGHUM-BASED BEERS AND THE STUDY ENVIRONMENT

1.1 SORGHO.

1.1.1 Origin and classification.

Sorghum is a species of monocotyledonous plant of the order Poale, in the family Poaceae (grasses), native to Africa (FAO, 1990). It is an annual herbaceous plant that can grow up to three metres high. It is grown either for its seeds (grain sorghum) or as fodder (forage sorghum).

1.1.2 Usage.

Sorghum is a cereal grown by people living in the semi-arid areas of Africa and Asia. It is a source of food for millions of people. Sorghum provides not only grain for human consumption, but also fodder (leaves, stalks), building materials, plastics, agrofuel and energy for cooking. (Barriere et al, 2003).

Sorghum has a wide range of uses in human nutrition, and its use at local level depends on the know-how of each people. In industry, it is used in the manufacture of many products. It is used in brewing as a complement or substitute for barley malt. Although its use is recent, it is increasingly used in the production of gluten-free beers (Bwanganga et al, 2013). Locally, sorghum is used in the manufacture of various products such as: flat bread, fermented flat bread, thick porridge, clear porridge (liquid) and beverages. Industrial uses include the production of biscuits, sugars, syrups, plastics and biofuels. Sorghum is used to make a variety of industrial products, including starch, alcohols, glucose, cellulose, paper, organic acids and a number of other chemical products useful in other industries.

1.1.3 Chemical composition and nutritional values of sorghum.

Table **1** presents information on the chemical composition and nutritional value of sorghum.

Table 1: Nutritional value of sorghum per 100g of whole grain

Nutrient	Unit	Quantity per 100g
Energy	Cal	386,02
Water	G	9,2
Protein	G	11,3
Lipid	G	3,3
Saturated fatty acids	G	0,46
Carbohydrate	G	68,33
Total sugars	G	68,33
Dietary fibre	G	6,3
Salt	G	0,02
Calcium	Mg	28
Iron	Mg	4,4
Phosphorus	Mg	287
Potassium	Mg	6
Vitamin B1	Mg	0,24
Vitamin B2	Mg	0,14
Vitamin B3	Mg	2,93
Monounsaturated fatty	G	0,99

acids		
Polyunsaturated fatty acids	G	1,37
Oleic acid	G	0,96
Linoleic acid	G	1,31
α-linolenic acid	G	0,07

Source: Canadian Sorghum Nutrient File, 2013

1.2. TRADITIONAL SORGHUM-BASED BEERS.

1.2.1. Definition

Traditional sorghum-based beers are beers traditionally obtained by alcoholic fermentation of a wort prepared from sorghum malt. These beers are produced in several African countries, but the production processes vary depending on their geographical location (François, 2013). They are very rich in calories, B vitamins including thiamine, folic acid, riboflavin and nicotinic acid, and essential amino acids such as lysine (Maoura, 2007).

1.2.2. Name

Traditional sorghum-based beers are widespread in several African regions and are known by a variety of names: "*Dolo*" in BOURKINA FASO, MALI and SENEGAL; "*Tchoukoutou*" in BENIN; "*Burukutu*" or "*Pito*" in NIGERIA and GHANA; "*Tchakpalo*" in NIGER and COTE D'IVOIRE; "*Bili bili*" in CHAD; "*Amgba*" in CAMEROON; "*Ikigage*" in RWANDA; "*Impéké*" in BURUNDI, "*Utshwala*" in SOUTH AFRICA; "*Umusemburo* or *Musururu*" in DR CONGO (Odoufa, 1985). Wherever it is made, '*musururu*' is drunk chilled.

"Musururu" generally has a dark brown colour, a pleasant aroma and a slightly acidic, sweet flavour. It is a beer prepared using a traditional process that mainly involves spontaneous fermentation of a wort obtained from sprouted sorghum and a large quantity of water (François, 2013).

1.2.3 Socio-economic importance

Musururu is highly appreciated by a segment of the Congolese population, where it plays an obvious social role. This beer is often served chilled in cabarets, which are popular bars and places for socialising (François, 2010). It is a traditional beer that is widely used during certain community celebrations and ceremonies. *Musururu"* is also sold almost everywhere in the streets and markets. For some consumers, '*musururu*' is an effective thirst-quencher and an excellent economic substitute for industrially-produced beers. It is an important source of income for rural populations (Baba-Moussa et al, 2012). Many brewers in both urban and rural areas derive most of their profits from this activity. Producing and selling '*musururu*' gives them a degree of financial independence. *Musururu*' is therefore of considerable socio-economic importance, and must be taken into account in any development action aimed at women in DR Congo (Konfo, 2016).

1.2.4. Production technology.

The traditional production processes for sorghum beers are long, complex and vary from country to country and region to region. However, they are based on the same principles. The three main operations are malting, mashing and fermentation (Chevassus-Agnès et al, 1976).

1.2.4.1. Malting

Malting is an operation which consists of making the grain germinate with the appearance of a shoot (FAO, 1995). It also refers to a process whereby sugars and an adequate proportion of enzymes for fermentation are obtained from the grain (Lasekan et al, 1995). In general, the diastatic power of malt varies with temperature, germination time and moisture content of the grain (Morral et al, 1995).

Malting generally involves three successive stages: steeping, germination and drying. According to Dirar (1978), 3 to 4 days of germination at a temperature ranging from 28 to 34°C after at least 22 hours of soaking the grains enable very good quality malt to be obtained from sorghum. During germination, the enzymes (α and β-amylases) produced in the grain germ spread into the endosperm and break down the starch into mono- and disaccharides (Hulse et al, 1980). The most active and abundant enzyme in sorghum malt is α-amylase with a β-amylase/α-amylase ratio ranging from 0.22/1 to 1/1 (Aisien, 1982). Drying slows down the activity of these enzymes, which are eventually inhibited once the milled, malted sorghum is placed in aqueous suspension (Dufour and Mélotte, 1992).

1.2.4.2. Brewing.

The aim of brewing is to solubilise as much of the water-soluble matter in the malt flour as possible to obtain a sweet mixture. Enzymes convert the starch in the malt and raw grains into simple sugars (glucose). Brewing consists of adding water to the malt flour (mashing) and heating the resulting mixtures at different temperature levels (Belliard, 2001).

1.2.4.3. Fermentation

Fermentation of cereals is a widespread practice in Africa (Hounhouigan, 1994). It can be defined as a process of biochemical modifications caused by micro-organisms and their enzymes on the products of first transformation (FAO, 1995). According to Eskin et al (1983), the conversion of monosaccharides (glucose, fructose) and disaccharides (lactose, maltose) into pyruvic acid mainly follows the glycolysis pathway. The subsequent route taken by pyruvic acid depends on the type of microorganism involved in the fermentation process and its metabolism: lactic acid bacteria convert pyruvic acid into lactic acid via a hydrogenation reaction. Yeasts, on the other hand, convert it into ethanol by a decarboxylation reaction (Leclerc et al, 1983). In the traditional *'musururu'* technological process, there is a double fermentation: alcoholic fermentation superimposed on natural lactic fermentation (François, 2010).

The following two types of lactic fermentation can be distinguished in the production of grain beers:

✓ Homolactic fermentation, in which the glucose is essentially transformed into lactic acid.

✓ Hetero-lactic fermentation where, in addition to lactic acid, other compounds are produced: carbon dioxide, acetic acid, ethanol and sometimes glycerol (Pasteur, 1995).

The products formed during fermentation are the results of the activation of enzymes secreted by the carbohydrate substrates, the homo and hetero fermentative bacteria involved in fermentation and the microflora present on the sorghum grains during production. Tables **2** and **3** show, respectively, the lactic acid bacteria, yeasts and moulds isolated from traditional sorghum beers in Africa.

Table 2: Species of lactic acid bacteria identified in traditional sorghum beers produced in sub-Saharan Africa.

Sorghum beer	Species of lactic acid bacteria	Identification methods	Reference
Chibuku (Zimbabwe)	*Lactobaciluus plantarum* *Lactobacillus delbrueckii* *Lactobacillus* sp	API 50 CH	Togo *et al*, 2002
Dolo and Pito (Burkinafa, Ghana)	*Lactobacillus fermentum* *Lactobacillus delbrueckii subsp.* *Bulgaricus* *Pediococcus acidilactici*	API 50 CH, ITS-PCR RFLP, PFGE, Sequencing of the 16S Rdna gene.	Sawadogo-Lingani *et al*, 2007
Ikigage (Rwanda)	*Lactobacillus fermentum* *Lactobacillus buchneri*	API 50 CH	François *et al*, 2010
Tchakpalo	*Lactobacillus fementum*	API 50 CH, PCR-	Djè *et al*, 2009

| (Ivory Coast | *Lactobacillus cellobiosus* | multiplex |
| | *Lactobacillus brevis* | |

Source: Coulibaly et al (2014)

Table 3: Yeast species identified in traditional sorghum beers produced in sub-Saharan Africa.

Sorghum beer	Yeast species	Identification methods	References
Pito (Ghana)	*Saccharomyces cerevisiae*		
	Candida tropicalis	Physiological and biochemical tests	Sefa-Dedeh *et al,* 1999
	Kloeckera apiculata		
	Hensenula anomala		
Bili-bili	*Saccharomyce cerevisiae*	PCR-RFLP of the NTS2 region, sequence of the D1/D2 domain of the 26S rDNA gene, CHEF	Nanadoum *et al,* 2005
	Kluyveromyces marxianus		
	Crptococcus albidius var. albidius		
	Candida melibiosica		
Tchakpalo (Ivory Coast)	*Saccharomyces cerevisiae*	PCR-RFLP of the ITS region, sequencing of the D1/D2 domain of the 26S rDNA gene	N'guessan *et al,* 2011
	Candida tropicalis		
	Pichia kudriavzevii		
	Kodamaea ohmeri		
Ikigage (Rwanda)	*Saccharomyce cerevisiae*	API 20 C AUX, sequencing of the ITS-5.8S rDNA	François *et al,* 2010
	Candida inconspicua		

	Issatchenkia orientalis	region	
	Candida magnolia		
Tchoukoutu **(Benin)**	*Saccharomyce cerevisiae*	API 20 C AUX	Kayodé *et al*, 2011
	Saccharomyces pastorianus		
	Candida albicans		
	Candida estchellsii		

Source: Coulibaly *et al* (2014).

1.2.5. Suitability for storing sorghum beers.

Most traditional African foods based on fermented cereals deteriorate rapidly and become unfit for consumption around four days after production. These changes are mainly due to the unpleasant aftertaste or strong acidification induced by the activity of microorganisms after production (Kutpacripo *et al*, 2009). Sorghum beers are no exception to this rule. The shelf life of sorghum beers has been reported as the main problem facing brewers in Sudan (Dirar, 1978), Tanzania (Tisekwa, 1989), Nigeria (Sanni *et al*, 1999), Rwanda (François *et al*, 2010) and Benin (Dossou *et al*, 2011). This beer is consumed while it is still fermenting.

The wort is not reheated or otherwise treated before the ferment is added, so the beer always contains residual microflora, mainly from its ingredients. Microorganisms such as *Staphylococcus aureus, Escherichia coli, Bacillus subtilis, Aspergillus niger, Aspergillus flavus, Rhizopus stalonifer, Saccharomyces cerevisiae* and certain species of *Streptococcus, Proteus* and

Mucor were isolated from the samples. The presence of these micro-organisms in these samples can be attributed to poor handling (failure to comply with hygiene rules) during production, which could be the cause of public health problems. Indeed, certain strains of *E. coli* can cause gastroenteritis, urinary tract infections and diarrhoea (Kolawolé *et al*, 2007). According to Baba-Moussa et al (2012), the environment and handling during manufacture and marketing are responsible for the contamination of these beers.

The authors added that it might be important to confirm these results by studying the microbiological quality of the raw materials used in its preparation before use. The metabolic activities of mesophilic lactic acid bacteria are also cited as the main culprits in the deterioration of these beers. These bacteria, along with other undesirable bacteria (Acetobacter), produce volatile acetic acid that has a bad taste and fruity odour, making the beer's taste, smell and texture unacceptable to consumers (François *et al*, 2013). The flash pasteurisation method increases the shelf life of industrial European beers by destroying harmful microbes. Unfortunately, this process is not applied to traditionally brewed sorghum beer. Early attempts at pasteurisation failed because they led to an unacceptable increase in the viscosity of the beer through further gelatinisation of the starch and elimination of the amylolytic enzymes. This operation also eliminated the characteristic effervescence of beer by destroying the active yeasts (Novellie *et al*, 1986). The results obtained by Dossou *et al*, (2011) show that sorghum beers fermented with yeast and the traditional commercial ferment and stabilised by pasteurisation at 60°C for one hour can be kept for at least 30 days at room temperature.

Table 4: Characteristics of some sorghum beers and suggested steps for improvement

Beer (Country)	Raw materials	Main phases	Suggested improvement steps	
Tchakpalo (**Benin/ Ivory Coast**)	Sorghum and/or maize grains	Malting, crushing, brewing, acidification, fermentation	-filtration, -pasteurisation (60°C/1h) -use of starter culture (*Saccharomyce cerevisiae).*	Aka et al, 2008, Dossou et al, 2011
Tchoukoutou (**Benin and Togo**)	Sorghum grain	Malting, crushing, mashing, brewing, acidification, fermentation	Stabilisation through the use of rustic wine technology	Osseyi et al, 2011
Bili-bili (**Chad**)	Sorghum grains	Malting, crushing, churning, brewing, acidification, fermentation.	Use of starter culture (*Saccharomyce serevisiae*)	Nanadoum et al, 2005
Ikigage	Sorghum grains	Malting, crushing, churning, brewing, fermentation	Brewing with *Vernonia amygdalina* leaf	François et al, 2013
Dolo (**Burkinafaso**)	Sorghum grain	Malting, crushing, churning, brewing, fermentation	Addition of *Hibiscus esculenta* juice	Traoré et al,2004.

Source: Konfo, (2016).

1.3. RESEARCH COMMUNITY

Our research was carried out in the town of Goma, and more specifically in the commune of Karisimnbi. The city of Goma is located south of the equator between 141° south latitude and 29°14 east longitude. (Mairie de Goma, 1999).

It is bordered to the north by Nyiragongo territory, to the south by South Kivu province and Lake Kivu, to the west by Masisi territory and to the east by the Republic of Rwanda. It covers an area of 66,824 km^2 or 11% of the province of North Kivu.

It is subdivided into communes, these into neighbourhoods and the latter into avenues and cells. The names and boundaries of the communes in the city of Goma are set out in order-law no. 89-127 of May 1989. These communes are

- ✓ The commune of Goma covers an area of 33.372 km^2 and has seven districts: Mikeno, Mapendo, Les Volcans, Katindo, Keshero, Himbi and Lac-Vert.
- ✓ The commune of Karimbi covers an area of 33.452 km^2 and comprises 12 districts: Kahembe, Murara, Bujovu, Majengo, Mabanga North, Mabanga South, Kasika, Katoyi, Nyabushongo, Ndosho, Mungunga and Virunga.

CHAPTER 2. RESEARCH MATERIALS AND METHODS

2.1. Hardware

The basic material used in this research consists of :

- ✓ a "*musururu*" drink obtained from a local brewer (in the Nyabushongo district, on 28 May 2017) (constituting sample 1)
- ✓ the two *musruru* beers produced in the laboratory with the addition of 30% sucrose (Sample2) and 25% sucrose (Sample3)

The laboratory experiments used the following equipment: culture media, Erlenmeyer, beaker, flasks, Petri dishes, pipettes, anaerobic jar, autoclave, incubator, microscope, slides, Gram staining reagents, balance, knife, sieves, grinder, bucket, pot, funnel, plastic bottles, plates, pH meter, test tubes, picnometer, alcoholometer, sterile syringes, alcohol lamp, paraffin lamp, cotton wool, stopwatch and reagents (NaOH, Phenolphalein).

2.2 Methods .

The following methodological approach was used to achieve our research objective:

- ✓ A survey was carried out among manufacturers and consumers of this alcoholic beverage to gather information on the various manufacturing techniques.
- ✓ Based on the recommendations of CE-ASEAN, 2005, the critical points in relation to the various unit operations in the manufacture of the said beverage were revealed.
- ✓ An improved beverage was produced after improving the critical points of the traditional technique.

✓ Finally, microbiological and physico-chemical analyses were carried out on the improved beverage compared with the traditional (unimproved) beverage.

2.2.1. Survey of local technologies for making the "*musururu*" drink

A survey on capitalising on the technological variants of "*musururu*" production in Goma was carried out in the commune of Goma, more specifically in the Nyabushongo district. The district was chosen because it is home to a large number of consumers and producers of the drink.

The methodology used in this research is that of a semi-structured survey based on an interview guide. The information contained in the interview guide related to personal details, raw materials used, production technologies (stages, equipment used and duration of operations) and maximum shelf life.

Given the lack of reliable information on the players in this sector, the respondents chosen were mainly brewers and consumers of "*musururu*" living in the selected district. Respondents were chosen at random on the basis of accessibility and their willingness to provide information. In all, 15 brewers and 10 consumers were interviewed.

2.2.2. Evaluation of critical points (ccp) in a manufacturing diagram

By definition, a critical control point (CCP) is a stage at which monitoring can be carried out and is essential to prevent or eliminate a health hazard or reduce it to an acceptable level (EC-ASEAN, 2005). Analysis must be carried out at each stage of the process. Three methods can be used to determine CCPs:

a) Use of decision trees

b) Applying the intuitive method

c) Application of hazard analysis to the production diagram

In the context of our research, hazard analysis based on the production diagram was used. To control the CCP, an operation-by-operation analysis must be carried out on the manufacturing diagram in order to avoid or eliminate any microbiological hazard, which can be of three types: contamination, multiplication and also the survival of microorganisms (Codex Alimentarius, 1999).

2.2.3. *Musururu"* production trial in the laboratory

After investigating the process of making traditional "*musururu*" beer at local level, we moved on to making it on a smaller scale (in the laboratory) in order to improve it. To do this, we proceeded as follows:

2.2.3.1.   Preparation and weighing of raw materials.

The sorghums of the "Muyimbudu" variety (the variety used as a raw material in the traditional manufacture of "*musururu")* were sorted and cleaned with tap water to remove any impurities (mud, cut grains, rotten grains, immature grains). The cleaning of the grains ended when the clear water began to flow.

2.2.3.2.   Malting.

Once all impurities had been removed, the sorghum grains were malted using the same techniques as those used by local brewers. We proceeded as follows: 5kg of sorghum grains were soaked in 5 litres of clean tap water (for 1 day). The grains were then removed from the water and ash was added. The seeds were then spread out on a bag and covered with another bag. Germination took 2 days, as did solar drying. The germinated grains were then separated from the ash and ground to produce flour.

2.2.3.3. Brewing.

This is the stage during which starch and proteins are hydrolysed by enzymes into fermentable sugars and nitrogen compounds. It is during this stage that the brew (Kikoma) is obtained. The brew was produced by brewing a single infusion. The brewing was carried out as follows:

- ✓ Cleaning, rinsing and disinfecting all containers to be used;
- ✓ Heat 6 litres of water to 45-55°C and add 3kg of malted sorghum flour while stirring, maintain this temperature for 20 minutes to allow protease production and then raise the temperature to 62-65°C and maintain this for 30 minutes to allow β-amylase production. After passing this temperature, the brew was brought to a temperature of 66-75°C for 30minutes to allow the production of α-amylase. The brew was then heated to 100°C for one hour to destroy all traces of microorganisms (sterilisation).
- ✓ The sterilised brew was then transferred to a series of containers to allow rapid cooling and avoid the risk of contamination.

2.2.3.4. Filtration.

The resulting brew was filtered to separate it from waste and other solid particles to obtain the wort that would be used for fermentation.

2.2.3.5.   Fermentation.

The must was fermented using the yeast *saccharomyces cereviseae*. The two treatments were carried out according to the quantity of sucrose used. To determine the amount of sucrose to be added to the must, we used the recommendation of Brun (2011), who said that "the amount of sucrose to be added to a must should correspond to 30% of the total weight of the must used and the yeast added should correspond to 3%". Based on this recommendation, we obtained a first treatment.

The second treatment was obtained by reducing the amount of sucrose to be added (25% instead of 30%) in order to assess the impact of this on the quality of the drink.

✓ **First treatment**

If the quantity of must is 2 L ⟶ 2kg ⟶ 2000g

We know that the quantity of sucrose to be added corresponds to 30% of the total weight of the must, i.e. 600g.

As the quantity of yeast to be added is 3% of the total weight of the must, we will have 60g to add.

✓ **Second treatment**

If the quantity of must is 2L ⟶ 2kg ⟶ 2000g

We will add a quantity of sucrose corresponding to 25% of the total weight of the must, i.e. 500g, and 3% of the yeast (i.e. 60g).

The fermentation device for each treatment was topped by a bubbling system to allow the carbon dioxide produced to escape. The disappearance of carbon dioxide (CO_2) in the bubbler made it possible to determine the duration of fermentation (which was 7 days).

2.2.3.6.   Packaging.

The beer obtained was racked, i.e. put into bottles (333 ml glass bottles) previously disinfected with ethyl alcohol and systematically rinsed with hot water. The beer was then pasteurised, heated in a water bath at 65°C for 1 hour to destroy all traces of yeast and micro-organisms. The bottles containing the beer were kept for 13 days in the laboratory at room temperature.

2.2.4 Physico-chemical analysis of drinks "*musururu*"
2.2.4.1. Collection of samples.

The 1-litre sample of the traditional beer "*musururu*" (E_1 This sample was labelled, placed in a cool box and sent to the laboratory of the Office Congolais de Contrôle (OCC/Goma) for various analyses. For the improved '*musururu*', samples were obtained after production in the laboratory (according to the method described in section **2.2.3.** of the second chapter), comprising respectively E_2 and E_3.

2.2.4.2. Determination of pH

The pH was determined with a potentiometric pH meter using a 50ml volume of sample placed in a beaker.

2.2.4.3. Determination of sugar content (°Brix) by refractometry.

The °Brix degree was determined using an ATAGO refractometer, using 1ml of the sample contained in a 10ml syringe.

2.2.4.4. Determination of volatile acidity expressed as acetic acid

The following procedure was used to determine volatile acidity.

- Take approximately 20ml of the sample and add 100ml of distilled water,

- Distil the mixture and then collect 50ml of the distillate

-Titrate with 0.1N NaOH until the colour changes in the presence of phenolphthalein as an indicator.

The results are expressed as % lactic acid equivalent, which is calculated by :

$$Ac = \frac{vb - 6,0}{V\ prise}\ en\ \frac{g}{l} \text{where :}$$

Vb= volume of NaOH ;

Vprise= volume of sample ;

2.2.4.5. Determination of fixed acidity expressed as tartaric acid

The following procedure was used to determine the tartaric acid content:

- Take 1ml of the drink and add 200ml of hot distilled water,

- Titrate with 0.1N NaOH in the presence of phenolphthalein as an indicator.

$$At = \frac{vb \times 7{,}50}{V\ prise}\ \ en\ \frac{g}{l}\ \text{where :}$$

Vb= volume of NaOH ;

Vprise= volume of sample ;

2.2.4.6. Determination of alcoholic strength.

The alcoholic strength was determined by Pycnometry. We proceeded as follows:

- ✓ Weigh 100g of test sample (Pe) into a distillation flask and add 50ml of distilled water.
- ✓ Distil and collect 100ml of distillate in a beaker containing 10ml of distilled water to avoid evaporation of the first drop of ethanol.
- ✓ Determine the density (d) of the distillate by Pycnometry.
- ✓ Finally, determine the alcoholic strength by volume from the table showing the specific weight (density) and the corresponding ethanol content at 20°C.

2.2.4.7. Determination of density.

To determine the density we used the picnometric method, measuring the weight of a precise volume of the samples and referring to the weight of the equivalent volume of distilled water.

$$d = \frac{(PV + \acute{e}chant.) - PV}{(PV + eau) - PV} = \frac{Masse\ de\ l'\acute{e}chantillon}{Masse\ de\ l'eau} = \frac{m}{m'}$$

PV= empty weight (g) ;

PV+ech= empty weight + sample (g) ;

PV + water= empty weight + water (ml)

2.2.5. Microbiological analysis.

The microbiological and hygienic quality of the samples was assessed by enumerating total aerobic mesophilic bacteria (TAM), yeasts and moulds, enterobacteria, lactic acid bacteria, total coliforms and faecal coliforms after inoculation on specific nutrient media. The different germs tested, the methodology used, the culture media used, their composition and preparation, and the incubation conditions are summarised in table **5** below.

Table 5: Composition and preparation of culture media used in research

Microorganisms	Growing environments	Composition	Preparation	Methodology used	Census
FAMT	*PCA (Plant count Agar)*	*For 1 litre of medium: Tryptone 5g, autolytic yeast extract 2.5g, glucose 1g, Agar Agar 15g*	*Suspend 23.5g of dehydrated powder in 1l of distilled or demineralised water. Bring to the boil slowly while stirring until completely dissolved [x] Dispense into tubes or flasks. Autoclave at 120°C for 15 minutes.*	*Pour PCA culture medium into the dishes. take 1ml of liquid from each sample and inoculate in PCA [x] medium. Incubate at 37°C for 24 hours*	*The count is carried out on Petri dishes containing 30 to 300 colonies.*
Total and faecal coliforms	*Mac Conckey broth*	*For 1 litre of medium: casein peptone 20g, lactose 10g, dried beef bile 5g, bromocresol purple 0.01g*	*Dissolve 35g completely in one litre of distilled water. After dividing 9ml into tubes fitted with inverted Durham tubes, sterilise by autoclaving (15 minutes at 121°C).*	*Pour the culture medium (Mac Conckey broth) into the tubes. Take 1ml of each sample and inoculate into the medium. Incubate at 37°C for 24 hours for total coliforms and at 44°C for 24 hours for*	*The presence of coliform bacteria was certified by acidification of the culture medium with or without gas production (positive tube) [xx]*

Yeast and mould	*Sabouraud dextrose agar*	*For 1 litre of medium: meat peptone 10g, glucose 35gr, agar agar 15g, pH= 5.7* [xx]	*Suspend 60g of dehydrated medium in one litre of distilled or demineralised water. Bring to the boil slowly, stirring until completely dissolved. Dispense into tubes or flasks.* [xxx] *Autoclave at 120°C for 15 minutes.*	*faecal coliforms.* *Pour the Sabouraud dextrose agar medium into the Petri dishes. Take 1 ml of each sample and inoculate into the medium. Incubate at 37°C for 48 hours.*	*The count is based on Petri dishes containing 30 to 300 colonies.*
Pathogenic Enterobacteriaceae (E.coli and Salmonella)	*Mac Conckey agar with crystal violet*	*For 1 litre of medium* [xx]*: casein peptone 17g, meat peptone 3g, lactose 10g, bile salt mixture 1.5g, sodium chloride 5g, neutral red 0.03g, crystal violet 0.001g, agar agar 13.5g*	*Suspend 50g in a litre of distilled water and leave to swell for 15 minutes. Heat until completely dissolved. Autoclave for 15 minutes at 121°C.*	*Pour the Mac Conckey crystal violet agar into the Petri dishes. Take 1ml of each sample and inoculate into the medium. Incubate at 37°C for 24 hours* [xxx]*.*	*The count is based on Petri dishes containing 30 to 300 colonies.*
Lactic acid bacteria	*Man, Rogasa and Sharpe (M.R.S.) agar*	*For 1 litre of* [xx] *medium: peptone 10g, meat extract 8g, yeast extract*	*Suspend 66g in a litre of distilled water and leave to swell for 15 minutes.*	*Pour the medium (M.R.S.) into the Petri dishes. Take 1 ml of each*	*The count is based on Petri dishes containing 30 to*

		4g, D(+) glucose 20g, bipotassium phosphate 2g, ammonium citrate 2g, sodium acetate 5g, $MgSO_4$ 0.2g $MnSO_4$ 0.05g, $FeSO_4$ 0.04g, tween 80 1ml, Agar 15g	Heat until completely dissolved. Autoclave for 15 minutes at 121°C.	sample and inoculate into the medium. Incubate at 37°C for 24 hours	300 colonies.

[x]Guiraud, 1998

[xx] Mercket, 1972.

[xxx] Difco, 1984

2.2.6. Statistical analysis.

The various analyses were carried out in three replicates in order to obtain a good average. Statistical analyses (analysis of variance at the 0.05 threshold and Student's T-test) were carried out using R2.13.1 software.

CHAPTER 3: RESULTS AND DISCUSSION OF THE RESEARCH.

3.1. PRODUCTION TECHNOLOGY AND CRITICAL POINTS IDENTIFIED DURING THE PRODUCTION OF THE TRADITIONAL DRINK "*MUSURURU*".

The results of the semi-structured survey provided information on the different technologies used by brewers to produce "*musururu*". It then enabled us to identify the various critical points in the production diagram for this drink.

3.1.1. Technological variations in *musururu* production.

At the end of this survey, we identified three different technological variants for the production of "*musururu*" in Goma. The preparation diagram for these different types of *'musururu'* is summarised in Figure 1.

✓ The traditional technological production diagram (**variant 1**) shown in Figure 1 is used by 20% of the brewers surveyed. This variant produces a beverage with a lower alcohol content than the other two techniques. For this beverage, fermentation is caused by the wild flora of yeasts developing on the sorghum grains before and after malting.

✓ The traditional technological production diagram (**variant 2**) shown in Figure 1 is used by 70% of the brewers surveyed. In this variant, apart from the wild flora growing on the sorghum grains, a ferment called "umusemburo or nshungo" (a quantity of the same drink prepared previously) is added to the mash before fermentation. The result is a highly alcoholic beverage that is much appreciated by consumers.

✓ The modified traditional technological diagram (**variant 3**) shown in Figure 1 and used by 10% of brewers. This variant is obtained by modifying variant

2, with the addition of honey to obtain a mead that is also highly alcoholic and very expensive compared to the two variants mentioned above.

Variant 2 was the subject of our research because it is widely consumed by a large number of customers and is slightly cheaper than variant **3**, the mead type. It is also appreciated for the fact that its taste is closer to the Western lager type beers produced by two companies in DR Congo (BRALIMA and BRASIMBA), so for consumers who do not have the means it is a good substitute beer par excellence for these Western beers.

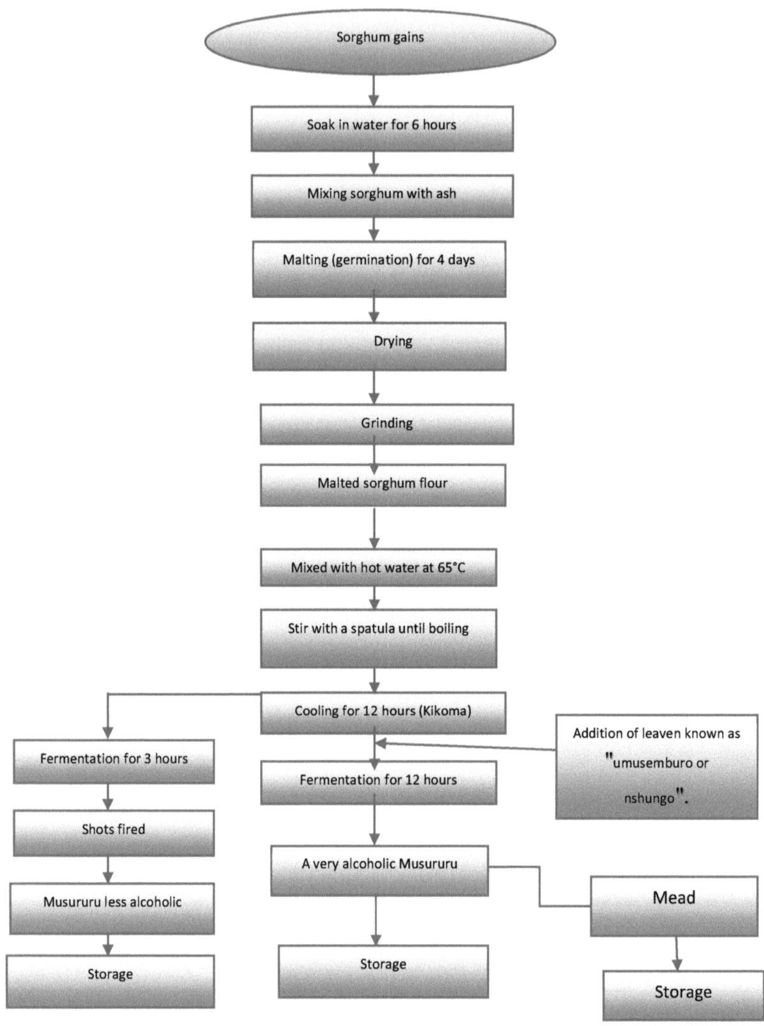

Variant 1 Variant 2 Variant 3

Figure 1Traditional technological diagrams for the production of "musururu".

3.1.2 Critical points identified during the production of "*musururu*".

After identifying the various '*musururu*' production diagrams at local level, a scrupulous analysis was carried out to determine and control the critical points.

These critical points are as follows:

➤ **Choice of raw materials:** the sorghum grains used are not dried in the field before harvesting, which leads to the development of certain spoilage micro-organisms.

According to Raemaekers (2001), the quality of the finished product depends on the raw materials used (cereals and water) and on the technology, as the processing of the grains affects their nutritional and chemical value.

This critical point can be controlled by using grains that have been dried in the field or at the brewery before the brewing process.

➤ **Steeping:** the duration of steeping is not well known; it is sometimes short and sometimes long, which makes it impossible to obtain the grain moisture content recommended for good malting. According to Ballegou (2014), to obtain a good malt, the steeping grains must reach a moisture content of 45% after one day's steeping. However, in the case of local brewers, this only lasts for 6 hours, which means that the grains cannot be effectively moistened for good malting. This critical point can be controlled by observing the 1-day steeping time recommended by Ballegou (2014).

➤ **Malting:** sometimes brewers do malting badly, the grains are not properly soaked, the malting time is not respected and this results in a lower rate of germinated grains, but the quality of the beer depends largely on the way in which malting has been done. According to Novellie et al, (1986), malting plays a major role in the beer-making process, and the final quality of the beer will depend on how this process is carried out, since it is at this stage that complex sugars are broken down into simple sugars. For these complex sugars to be broken down into simple sugars, malting must last 7 days (Novellie et al, 1986). The 7-day malting period is therefore necessary to control this critical point.

➤ **Drying:** the substrates used are not clean, and as drying is traditional (in the sun), it is carried out outdoors in a poorly maintained environment, exposing the product to the weather, microbes and predators, leading to risks of prevalence of contaminants and especially mycotoxins (aflatoxins, fumonisins etc.....). Cleaning the surfaces to be used before and after each use and maintaining the drying area are essential to control this critical point.

➤ **Grinding:** the mills used to grind the grains are also used for various other products (cereals, vegetables, etc.....). This results in sorghum flour being mixed with other products, which has a negative impact on the quality of the finished product. This critical point can be controlled by using mills specialised solely in sorghum grinding.

➤ **Brewing:** this stage is critical because brewing machines do not respect the time/temperature pairing for the production of the various enzymes. This results in a brew that is low in fermentable sugar. In addition, the brew is not boiled long enough to kill off all the micro-organisms in the brew. In order to control this critical point, it is necessary to respect the temperature stages for the production of the various enzymes, as well as a suitable time-temperature scale for the sterilisation of the brew.

➢ **Fermentation:** this is a critical point in the production of "*musururu*", as it is carried out using a leaven called "umusemburo", which is obtained traditionally and whose true nature is unknown. Brewers sometimes even use the *musururu* from the previous fermentation as a ferment.

According to Clerck (1980), true alcoholic fermentation must be carried out by a ferment selected for its fermentative and aromatic qualities (*saccharomyce cereviseae*). Using the selected *saccharomyce cereviseae* ferment instead of the traditional 'umusemburo' ferment can help to control this critical point and thus improve the quality of the drink.

➢ **Storage:** storage is a critical point because once the *musururu* has been produced, it is stored in containers that are not properly cleaned, rinsed and disinfected. Cleaning and disinfecting storage containers is necessary to control this critical point.

After determining the critical points on the diagrams of different technological variants for the production of "*musururu*", we proposed a new production technique using *Saccharomyce cerevisiae* yeast as the fermentation agent. We have thus presented a new diagram of improved production, shown in Figure **2** below.

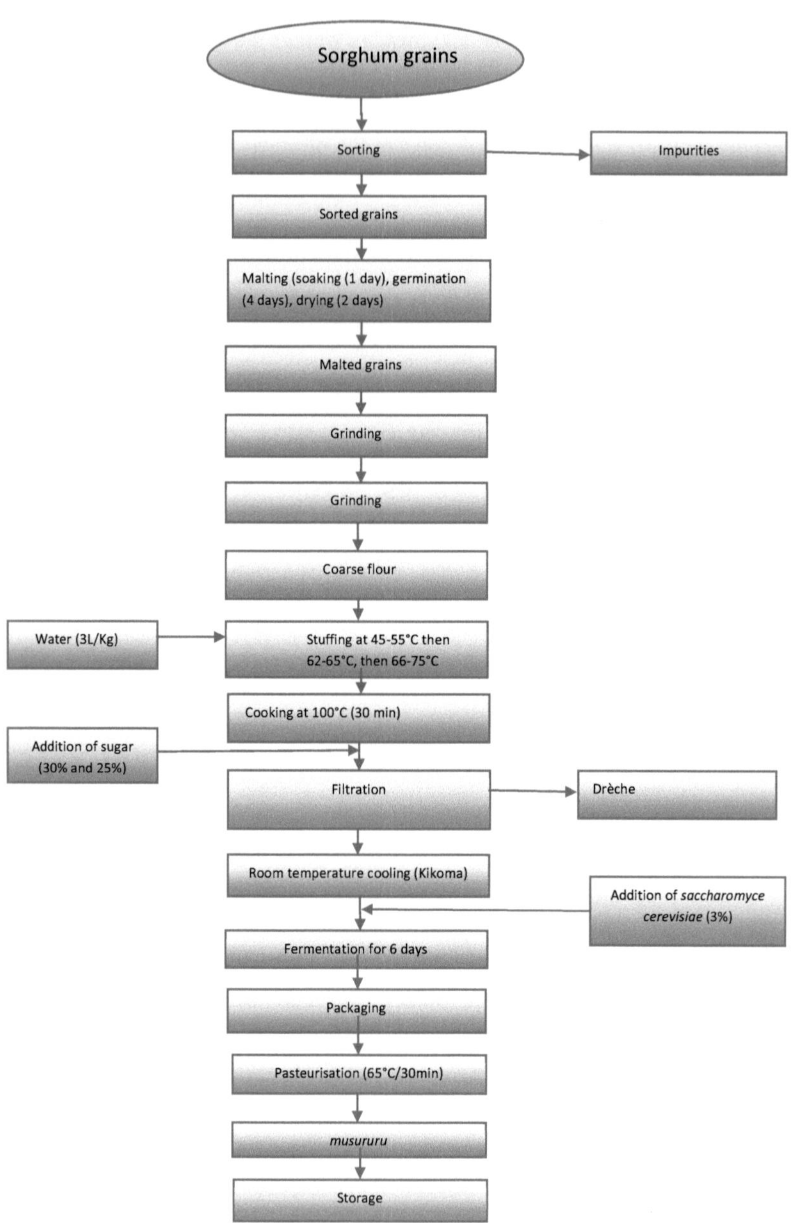

Sorghum grains
Sorting
Impurities
Sorted grains
Malting (soaking (1 day), germination (4 days), drying (2 days)
Malted grains
Grinding
Grinding
Coarse flour
Water (3L/Kg)
Stuffing at 45-55°C then 62-65°C, then 66-75°C
Cooking at 100°C (30 min)
Addition of sugar (30% and 25%)
Filtration
Drèche
Room temperature cooling (Kikoma)
Addition of *saccharomyce cerevisiae* (3%)
Fermentation for 6 days
Packaging
Pasteurisation (65°C/30min)
musururu
Storage

Figure 2Diagram of how "musururu" is made after the critical points have been improved

3.2 Assessment of physico-chemical parameters during the preservation of local and improved "*musururu*" drinks.

The physico-chemical parameters were assessed on the various "musururu" samples. These were the beverage harvested from a local manufacturer using technology **no. 2** in figure **1** (E_1) and two improved beverages (E_2 & E_3) as shown in Figure 2

Table 6 below shows the changes in various physico-chemical parameters of each beverage at the initial time (**t=0**, just after fermentation) and after **13** days of storage. An additional table (**see appendix 2**) compares the averages for each parameter for three samples over a given period.

Table 6: Evaluation of various physico-chemical parameters of each drink according to storage time.

Samples	Time (days)	Ph	Density	Alcohol content (% vol)	Acetic acidity (g/l)	Tartaric acidity (g/l)	Brix
E_1	00	3,804±0,147a	2,003±0,13a	3,97±0,11a	0,91±0,02a	6,4±0,45a	10,4±0,7a
	13	3,002±0,14b	2,002±0,14a	3,02±0,14b	1,68±0,9b	5,02±0,15b	10,4±0,7a
E_2	00	4,009±0,147a	1,904±0,164a	3,77±0,11a	1,3±0,28a	6,36±0,081a	12,78±0,72a
	13	4,006±0,057a	1,903±0,12a	3,76±0,15a	1,33±0,27a	6,34±0,329a	12,77±0,09a
E_3	00	4,923±0,25a	1,804±0,15a	4,08±0,11a	1,27±0,02a	6,99±0,89a	13±1,41a
	13	4,925±0,22a	1,801±0,12a	4,09±0,31a	1,26±0,06a	6,98±0,35a	13±1,21a

Mean values followed by the same letter in the same column and for the same sample are not significantly different at the 5% threshold. E_1 : Local "Musururu" obtained from the manufacturer; E_2 : Musururu obtained by adding 3% Saccharomyces serevisiae yeast, 30% sucrose and then stabilised by normal pasteurisation; E_3 : Musururu obtained by adding 3% Saccharomyce serevisiae yeast, 25% sucrose and then stabilised by normal pasteurisation.

Table **6** shows that for the sample E_1 the pH, alcohol content, acetic acidity and tartaric acidity changed during storage, as the statistical analyses show significant differences. For samples E_2 *and* E_3, samples, no significant differences were observed during storage. A comparison between the different beverages, at a given time (table **6**), shows that the control sample (E_1) obtained from the manufacturer showed poor keeping qualities. For this sample E_1, the alcohol content fell from 3.97±0.11 to 3.02±0.14 after 13 days of storage. This decrease is due to the fact that an alcoholic beverage after fermentation undergoes transformations if it has not undergone any stabilisation treatment (Konfo, 2016). For example, the ethanol contained in beer is transformed into acetic acid (Renouf, 2013). However, despite this reduction, the alcohol content remained within the two recommended limits for alcoholic beer, which vary between 3.5 and 7% vol (Andrews et al 2013). Also, the pH decreased from 3.804±0.147 to 3.002±0.14 during storage. This phenomenon is easy to explain if we link it to the alcohol level. If the alcohol is transformed into acetic acid, it is clear that the pH of the medium decreases. As the alcohol is transformed, acidity, especially volatile acid, increases and the pH of the medium decreases (Hervé et al, 2008). This explains the increase in acetic acid from 0.91±0.02 to 1.68±0.29. This phenomenon reinforces the explanation given for the above phenomena (decrease in alcohol content and pH). As ethanol oxidises (CH_3CH_2OH) is transformed into acetic acid (CH_3COOH), which is a volatile acid (Ndamanisha and Nkuririmana, 2016).

Tartaric acid decreased from 6.4±0.45 to 5.02±0.15. This decrease can be explained by the fact that tartaric acid is transformed into acetic acid by the action of lactic acid bacteria of the genus lactobacillus (Herve et al, 2008).

It should be noted that for the samples (*E_2 and E_3*) (obtained by adding sugar, fermentation by *Saccharomyces* and pasteurisation), there was no considerable variation in the parameters during storage.

All these parameters remained the same from the beginning to 13^e days of storage. We can therefore say that pasteurisation seems to stabilise all the parameters we are looking for in *"musururu"* beers.

3.3 Evaluation of microbiological parameters during the preservation of local and improved *"musururu"* beverages.

The following germs: total aerobic mesophilic flora, *Salmonella, E.coli,* faecal coliforms, total coliforms, lactic bacteria, yeasts and moulds were identified or detected in the three different *"musururu"* samples

Table **7** below shows the evolution of different germs in each beverage during storage. An additional table (see appendix **3**) shows a comparison between the averages for each germ in three samples over a given period.

Table 7: Evolution of different germs in each beverage during storage

Sample	Time	FAMT (UFC)/ml	Salmonella (CFU/ml)	E.coli (CFU/ml)	Faecal coliforms per ml	Total coliforms per ml	Lactic acid bacteria (CFU/ml)	Yeasts and moulds (CFU/ml)
E_1	00	1970±70,7a	00±0,0	00±0,0	00±0,0	00±0,0	1720±31,11a	2030±7,07a
	13	2360±70,7b	00±0,0	00±0,0	00±0,0	00±0,0	2060±21,21b	2940±28,28b
E_2	00	470±28,28a	00±0,0	00±0,0	00±0,0	00±0,0	00±0,0	370±14,14a
	13	630±21,21b	00±0,0	00±0,0	00±0,0	00±0,0	00±0,0	410±14,14b
E_3	00	450±14,14a	00±0,0	00±0,0	00±0,0	00±0,0	00±0,0	370±14,14a
	13	475±7,07b	00±0,0	00±0,0	00±0,0	00±0,0	00±0,0	390±21,21b

E_1 : Musururu obtained from the manufacturer, E_1 : Musururu obtained by adding 3% of the yeast Saccharomyces cerevisiae, 30% sucrose then stabilised by normal pasteurisation; E_3 : Musururu obtained by adding 3% of the yeast Saccharomyce cerevisiae, 25% sucrose then stabilised by normal pasteurisation; FAMT: Total aerobic mesophilic flora, CFU: Colony forming unit. Mean values followed by the same letter in the same column and for the same sample are not significantly different at a threshold of 5%.

Table **7** shows that *E.coli, Salmonella,* faecal coliforms and total coliforms were absent from the various samples. The absence of these two pathogenic germs (*E.coli, Salmonella*) is not surprising in '*musururu*' drinks, as this is not a good environment for their development. Because of its acidity, anaerobiosis and pressure, beer is not a suitable medium for the development of pathogens, which die quickly (Mol, 1991). The samples E_2 *and* E_3 are free of lactic acid bacteria, whereas the sample E_1 contains them and the quantity changes over time. For the other germs that remain present in the three samples (FAMT, lactic bacteria, yeasts and moulds), their quantity also changes during storage.

The gradual acidification of the '*musururu*' drink encourages the natural elimination of pathogenic germs from the product, so that after a few days of production, a healthier drink is obtained from a highly contaminated base product (Soma, 2014). This absence of pathogens (*E.coli, Salmonella*) and flora indicative of faecal pollution (faecal and total coliforms) gives the product an acceptable quality.

The total absence of lactic acid bacteria in the samples (E_2 *and* E_3) is due to the fact that these samples have been pasteurised. This is in agreement with Nout (2003), who states that pasteurisation leads to selective destruction of the microbial flora present in the food.

On the other hand, changes in the total aerobic mesophilic flora, yeasts and moulds and lactic acid bacteria were observed in the different fractions during storage for the sample (E_1) obtained from the manufacturers, the values for total aerobic mesophilic flora increased from 1970 ± 70.7 to 2360 ± 70.7. In samples stabilised by pasteurisation, the total aerobic mesophilic flora increased from 470±28.28 to 630±21.21 for sample (E_2) and from 450±14 ,14 to 475±7,07 for sample (E_3).

Very little work has been done on *musururu* in the scientific literature. There are virtually no specific standards for this local product. As '*musururu*' is an alcoholic beverage, these values can be compared with existing quality standards for beverages in this category. There is not always a close relationship between a high total aerobic mesophilic flora value and the presence of pathogenic microorganisms (Ouathara, 2010). However, as an index of sanitary quality, a risk to consumer health is considered to exist only if the total aerobic mesophilic flora is greater than 10^4 cfu/ml or g (Quebec, 2015). In principle, a total aerobic mesophilic flora exceeding 10^5 à 10^6 cfu/ml or g of product causes product deterioration (marketability index) (Quebec, 2015). This leads us to say that the two samples stabilised by pasteurisation comply with the standards and even the sample obtained from the manufacturers, despite their high concentrations of total mesophilic aerobic flora.

The change in lactic acid bacteria in the control sample (E_1), which rose from 1720±31.11 to 2060±21.216, this can be explained by the fact that lactic acid bacteria are introduced by the environment and the equipment used. Some authors mention the presence of lactic acid bacteria in traditional African cereal-based drinks (Coulybali, 2013). These bacteria are thought to be part of the ferment's natural environment. These lactic acid bacteria are involved in alcoholic fermentation and in the significant hydrolysis of non-fermentable sugars (Aka et al, 2008). The consequence of this consumption of sugars is a reduction in the dry extract and density of the '*musururu*'.

Yeasts and moulds in the control sample rose from 2030±7.07 to 2940±28.28. The presence of yeasts and moulds in large numbers in this sample is undoubtedly due to the fact that these microorganisms are brought in by natural fermentation and their evolution is not surprising since this sample did not undergo any stabilisation treatment. On the other hand, the presence and development of yeasts and moulds in samples stabilised by pasteurisation *(E$_2$ and E$_3$)*, which rose from 370±14.14 to 410±14.14 for the sample E_2 and from 370±14.14 to 390±21.21 for sample E_3 is surprising because it does not reflect the results found for the parameters (pH and alcohol content) which did not change during the 13 days of storage. In addition, lactic acid bacteria were absent after pasteurisation and even after 13 days of storage. This observation warrants further investigation.

CONCLUSION AND OUTLOOK

The aim of this research was to improve the traditional drink "*musururu*", which is made by hand.

The survey carried out in the course of this research enabled us to identify three different technological variants for the production of "*musururu*" and to identify the critical points associated with each unitary manufacturing operation. This enabled us to prepare a drink under controlled conditions.

Physico-chemical and microbiological analyses were carried out on "*musururu*" obtained from manufacturers and those developed by ourselves.

The technological variants of "*musururu*" production, compared with modern brewing technologies, are distinguished by the greater complexity of the brewing operation, the speed and brevity of alcoholic fermentation by the traditional ferment, and the superimposition of lactic fermentation.

Evaluation of the physico-chemical and microbiological characteristics of "*musururu*" obtained from local manufacturers and those made by ourselves (then pasteurised), reveals that the control sample E_1 showed significant variability (P< 0.05) in pH, alcoholic strength, acetic acidity and tartaric acidity during storage.

Samples stabilised by pasteurisation appear to be more stable than the control.

This same trend was observed for the microbiological analyses for the control and revealed significant variability (P<0.05) in FAMT, lactic acid bacteria and yeasts and moulds.

In view of the results obtained in the present research, we suggest a plan for the development and practice of pasteurisation for the preservation of this beverage in order to limit the process of alteration, infection and intoxication due to microorganisms installed in the finished product. This can easily be done, even by local manufacturers.

In order to enhance the value of this popular drink, we could try to make a pro-biotic   *"musururu* ". To do this, we need to master all the phenomena that take place in the beverage during fermentation, phenomena arising from the metabolisms of different groups of micro-organisms. Lactobacillus, with its many technological properties, should be of particular interest.

BIBLIOGRAPHY

1. Aka, s, Djeni, N.T, N'sugessan, K.F., Yao, K.C and DJP, K. (2008): Variability of physico-chemical properties and enumeration of fermentative flora of tchapolo a traditional sorghum beer in Côte d'Ivoire. Afrique science. International Journal of Science and Technology.

2. Alais et al, 1991 : *Abrégé de biochimie alimentaire.* 4th edition ; Paris : Maisson- 160-165

3. Andrews et al, 2013. Study of proposed changes to the Canadian beer standard of identity. P22

4. Ainsien, A.O, 1982. Enzymic modification of sorghum endosperm during seedling growth and malting. Journal of science of food and Agriculture, 33(8) 754-759.

5. Ballegou, 2014: Improving the traditional sorghum (sorghum bicolor) malting process for chakpalo production. Afrique science. International journal of science and technology.

6. Barriere et al, (2003). Growing dehusked sorghum in West and Central Africa, current situation and definition of a regional action plan.

7. Baba-Moussa, Adjanhoun, A., Attakpa, Kpavde, Gbenou, K., Simeon, 2012. Antimicrobial activity of three essential oils from Benin on five oral germs: Test of mouthbaths. Annals of Biological Research. 3(11). 5192-5199

8. Bourgois, C., 1998, *La biere et la brasserie,* Presses Universitaires de France, Paris, P126

9. Belliard, F (2001). La préparation de bière de sorgho chez les joohe (Burkina-Faso), étude ethnolinguistique d'une technique. Journal des africanistes.

10. Bwanganga. T, Khady, Ba, Jacqueline. D, Poul. M, Françoi B. (2013). Towards the integration of sorghum as a raw material for modern brewing. Université de liège Gemblaux Agro-Biotechnologie. (Belgium).

11. CE-ASEAN, 2005: guidelines on HACCP, good manufacturing practice and good hygiene practice for small and medium-sized enterprises.

12. Clerk, 1980 "Brewing Courses: Raw Materials, Manufacture and Installation. 2nd edition, vol 1.

13. Coulibaly, (2013). Etude comparative de procède traditionnels de préparation d'une bière appelée " tchapalo " 2ème journée annuelle de la SOACHIM, DAKAR-Ségal.

14. Coulibaly, W. H, N'guessan, K.F, Coulibaly.I. Djè K.M and Thonart. P (2014). Yeasts and lactic acid bacteria involved in traditional sorghum-based beer produced in sub-Saharan Africa. (Bibliographic synthesis. Biotechnology, Agronomy, Society and Environment).

15. Codex Alimentarius: CAC/RCP 1-1969, Rev3 (1997), Amended in 19991.

16. Chevassus-Agnes et al, 1979. Traditional technology and nutritive value of comeroon sorghum beers. Cah onerest, 2(1), 83-192.

17. Dicko, M.H., Guppen, H, Traoré, A.S., Van Berkel, W.S, and Voragen. A. G. (2005). Evaluation of the effect of germination on phendic compounds and antioxidant activates in sorghum varieties. Journal of Agricultural and food chemistry.

18. Difco, 1984: Difcco Manual: Deshydrated culture media and reagents for microbiology. Tenthedition; Difco Laboratories; Michigan; USA; 1155 pages.

19. Dirar, H. A. (1978). A microbiological of judanese merisa brewing. Journal of food science.

20. Dossou, J. Ballogou, V and Desouza, C. (2011). Comparative study of microbial dynamics and quality of chakpalo fermented with commercial yeast (*Saccharomyces cerevisiea*) and traditional ferment and stabilized by pasteurization. Journal of scientific research of the university of lome.

21. Dufour, S.P and Mellotte, 1992. Sorghum malts for the production of a lager beer. Journal of the American Society of Brewing Chemists. 50(3). 110-119

22. Eskin et al, (1983). Use of *Sacchoromyces* cerevisieae in the production of bili-bili, a local sorghum and maize beer. Proposal for improvement.

23. FAO, (1990) : Alimentation des aliments tropicaux céréales ; collection ; FAO : Alimentation et nutrition numéro 832

24. FAO, (1995). Sorghum and millet in human nutrition. Rome.

25. Canadian File, (2013). Chemical composition of *Sorghum* bicolor from Canada: Effects of drying and harvest site.

26. François, 2013: Characterisation and quality improvement of the traditional Rwandan beer "Ikigage" made from sorghum. Université de liège, Gemloux Agro-Biotechnologie, (Belgium), 208p

27. Francois, Kamaliza, G., Bajyana, E et Thanart, P.H, 2010. Microbiological and physico-chemical characteristic of Rwandese beer "Ikigage". African Journal of Biotechnology, 9(27). 4241-4246.

28. Guiraud, 1998: *Microbiologie* alimentaire, Editions Dunod, Paris, 196p

29. Herve et al, 2008. Les bactéries lactiques en œnologie, Lavoisier /Tec et doc 172 p.

30. Hounhouigan, (1994). Characterisation and frequency distribution of species of lactic acid bacteria involved in the processing of mawe, a fermented maize dough from Benin, Int. J. Food microbial (18). 279-287.

31. Hulse. J.H, Laing, E. M. and Pearson, O.E (1980). Sorghum and the millets, their compositions and nutritive value. Academic press. 997pp.

32. Kolawole, Houhouigan, D.J, Nout, M.J and Nehof, A, 2007. Household production of sorghum beer in Benin: technology and socioeconomic aspects. International Journal of studie, 31(3). 258-264.

33. Konfo (2016). Endogenous production practices and effectiveness of aromatic plant extracts in the stabilization of "Tchakpalo" a sorghum beer, in Benin. University of Abomey-Calava (Benin).

34. Kutpacripo, J., Parawira, W., Tinofa, J.S Koudita, I. and Ndengu, C. (2009). Investigation of shelf-life extension of sorghum beer (chibuku) by removing the second conversion of malt. International journal of food microbiology.

35. Laseken et al, 1995. Effect of germination and degree of gring (coarse/fine) on the extract and sugar contents of sorghum malts. Food Chemistry, 53(2). 125-128.

36. Leclerc, Hussan and Jakubizar, 1983. Microbiologie général, 2e édition Doin Editeurs, Paris. 287P

37. Louise A, Poul K, Akanza and Marboua(2007). Growing sorghum well in sub-Saharan Africa. Afrique science. International journal of science and technology.

38. Mairie de Goma, (2002). La problématique du lotissement dans la ville de Goma. Cas des quartiers Keshero et Katoyi.

39. Maoura, Mbailou, claude and jacque (2007). Analytical and microbiological technical monitoring of "bilibili", a traditional Chadian beer. Afrique science. International journal of science and technology.

40. Marcket, 1972: *Manuel de microbiologie*, Dermestadt, Germany, 445p

41. Moll, T., Tebb, G., Surana, Robistsch and Nasmyth, 1991. The role of phosphorylation and the CDC28 protein Kinase in cell cycle-regulated nuclear import of the *S.* cerevisiae transcription factor 8w15. Cell, 66(4), 743-758.

42. Morrall, P., Boyd. H. K., Taylor, J.R.N and Van Der Walt. W. (1995). Effect of germination and degree of grand (ecourse, fine) on the extrat and sugar contens of sorghum malt. Rood chemistry.

43. N'da. A. et al 1996: Comparative study of traditional processes for preparing a local beer called tchapolo.

44. Ndamanisha J. C and Nkuririmana J. (2016). Effect of local banana wine consumption on uric and elimination: a quantitative analysis. African journal of science and Research.

45. Nout R. (2003). *Food, processing, preservation and quality*. Afrique science. International journal of science and technology.

46. Novellie, L and Schaepdryver. P 1986. Modern developments in traditional African beers progress in industrial microbiology,

47. Odoufa, (1985). African fermented food in wood B.J.B, ed. Microbiology of fermented foods. London; New York, USA: Elsevier Applied Science Publishes, 155-191.

48. Pasteur L. (1995). Memoir on lactic acid fermentation (Extracted by the author). Molecular Medicine. 59p

49. Quattara C.T.A (2010). Cours de microbiologie alimentaire. Ufr-svt. Université d'Ouagadougu. 67p.

50. Quebec,(2015). *Microbiological criteria applicable to foodstuffs, guidelines for interpretation*. The Government of the Grand Duchy of Luxembourg.

51. Renouf, (2003). Les fermentations malolactiques dans les vins, Mecanismes et applications pratiques, Lovoisier/ Tec et Doc. 223P

52. Raemaekers R, (2001). Agriculture en Afirique tropicale ; DGO ; Bruxeles ; Belgique, P.1634.

53. Sanni, Onilude, A. A., Fadahunsi and Afolabi, 1999. Microbiological determination of traditional alcoholic beverage in Nigeria. Food research international.

54. Soma (2014). Use of lactobacillus fermentum culture in the technology of Zoom-Komm, a local drink based on millet (Penisetum Glaucum) to improve its nutritional, sanitary and organoleptic quality. One-line dissertation.

55. Tisekwa, B, (1989). Improvement of traditional manufacturing of sorghum beer (matama) in Tanzania: verbeting van het traditionele productic process van sorghum beer (mtama) in Tanzania (Doctoral dissertation, Gent: Risksuniversiteit Gent). 225P

APPENDIX

Date and place of survey

INFORMATION ON THE SURVEY

Name :

Age :

Sex: ....F......................... M............F....

RAW MATERIALS USED :

What types of raw materials do you use?

a. Sorghum

b. Maize + Sorghum

c. Sorghum + Milet

d. Sorghum+Maize+Millet

Do you add anything to the must before fermentation?

a. Sugar

b. Honey.

c. None.

Do you add anything to the drink after fermentation :

a. Sugar

b. Honey

c. No

PRODUCTION TECHNOLOGY.

Equipment used: what type of equipment do you use to produce musururu?

    a. Plastic material.

    b. Metal material

Production stage and duration of operations.

Maltage..Pendant :

Soaking...Pendant :

Germination..Pendant :

Drying...Pendant :

Brewing...Pendant :

Fermentation...Pendant :

What kind of ferment do you use to make Musururu?

    a. Saccharomyce serevisieae

    b. Lactobacillus fermentum

    c. Saccharomyce serevisieae var elliptical

    d. Saccharomyce uvarum

    e. Traditional ferment (Umusemburo).

How long does musururu ferment?

    a.12h

    b. 24h

c. 48 hours

d. 72h.

What is the shelf life of musururu :

a. 1 day
b. 2 days
c. 3 days

d. 7 days

e. 14 days

f. 21 days

Appendix 2. Table showing a comparison between the averages, for each parameter, of three samples for a given period

Samples	Time (days)	pH	Density	Alcohol content (% vol)	Acetic acidity (g/l)	Tartaric acidity (g/l)	Brix
E_1	00	3,804±0,147a	2,003±0,13a	3,97±0,11a	0,9±0,02a	6,4±0,45a	10,4±0,7a
	13	3,002±0,14b	2,002±0,14a	3,02±0,14b	1,68±0,9b	5,02±0,15b	10,4±0,7a
		Average : 3,403	Average : 2,0025	Average : 3,495	Average : 1,29	Average : 5,71	Average : 10,4
E_2	00	4,009±0,147a	1,904±0,164a	3,77±0,11a	1,3±0,28a	6,36±0,081a	12,78±0,72a
	13	4,006±0,057a	1,903±0,12a	3,76±0,15a	1,33±0,27a	6,34±0,329a	12,77±0,09a
		Average : 4,0075	Average : 1,9035	Average : 3,665	Average : 1,315	Average : 6,35	Average : 12,775
E_3	00	4,923±0,25a	1,804±0,15a	4,08±0,11a	1,27±0,02a	6,99±0,89a	13±1,41a
	13	4,925±0,22a	1,801±0,12a	4,09±0,31a	1,26±0,06a	6,98±0,35a	13±1,21a
		Average : 4,924	Average : 1,8025	Average : 4,085	Average : 1,265	Average : 6,985	Average : 13

Appendix 3 A table showing a comparison between the averages, for each germ, of three samples over a given period.

Sample	Time	FAMT (UFC)/ml	*Salmonella* (CFU/ml)	*E.coli* (CFU/ml)	Faecal coliform per ml	Total coliforms per ml	Lactic acid bacteria (CFU/ml)	Yeast and mould (CFU/ml)
E_1	00	1970±70,7a	00±0,0	00±0,0	00±0,0	00±0,0	1720±31,11a	2030±7,07a
	13	2360±70,7b	00±0,0	00±0,0	00±0,0	00±0,0	2060±21,21b	2940±28,28b
		Average : 2,165	Average : 00	Average : 00	Average : 00	Average : 00	Average : 1,890	Average : 2485
E_2	00	470±28,28a	00±0,0	00±0,0	00±0,0	00±0,0	00±0,0	370±14,14a
	13	630±21,21b	00±0,0	00±0,0	00±0,0	00±0,0	00±0,0	410±14,14b
		Average : 550	Average : 00	Average : 00	Average : 00	Average : 00	Average : 00	Average : 390
E_3	00	450±14,14a	00±0,0	00±0,0	00±0,0	00±0,0	00±0,0	370±14,14a
	13	475±7,07b	00±0,0	00±0,0	00±0,0	00±0,0	00±0,0	390±21,21b
		Average : 462,5	Average : 00	Average : 00	Average : 00	Average : 00	Average : 00	Average : 380